Contents

NATIONAL BIRD
STATE BIRD
STATE GAME BIRD
STATE RAPTOR
STATE DUCK
STATE PEACE SYMBOL

Welcome to Feathered & Famous!

Did you know that every state in the United States, and even the nation's capital, Washington D.C., has chosen at least one bird to represent it? It's true! These birds are like team mascots. Each one symbolizes something unique and special about the place it calls home.

This book is your invitation to meet America's Bird Hall of Fame and discover what makes each of these birds extraordinary.

But we're not stopping at State Birds. We're going even bigger! You'll meet other official birds such as State Ducks, State Raptors (birds of prey), State Game Birds (hunted in season), and even a bird that represents peace. And let's not forget the most famous of all: the National Bird of the United States, the Bald Eagle.

Get ready to take flight and meet the birds with special honors in America. Some sing, some soar, some swoop, but every single one has a story to tell!

United States of America

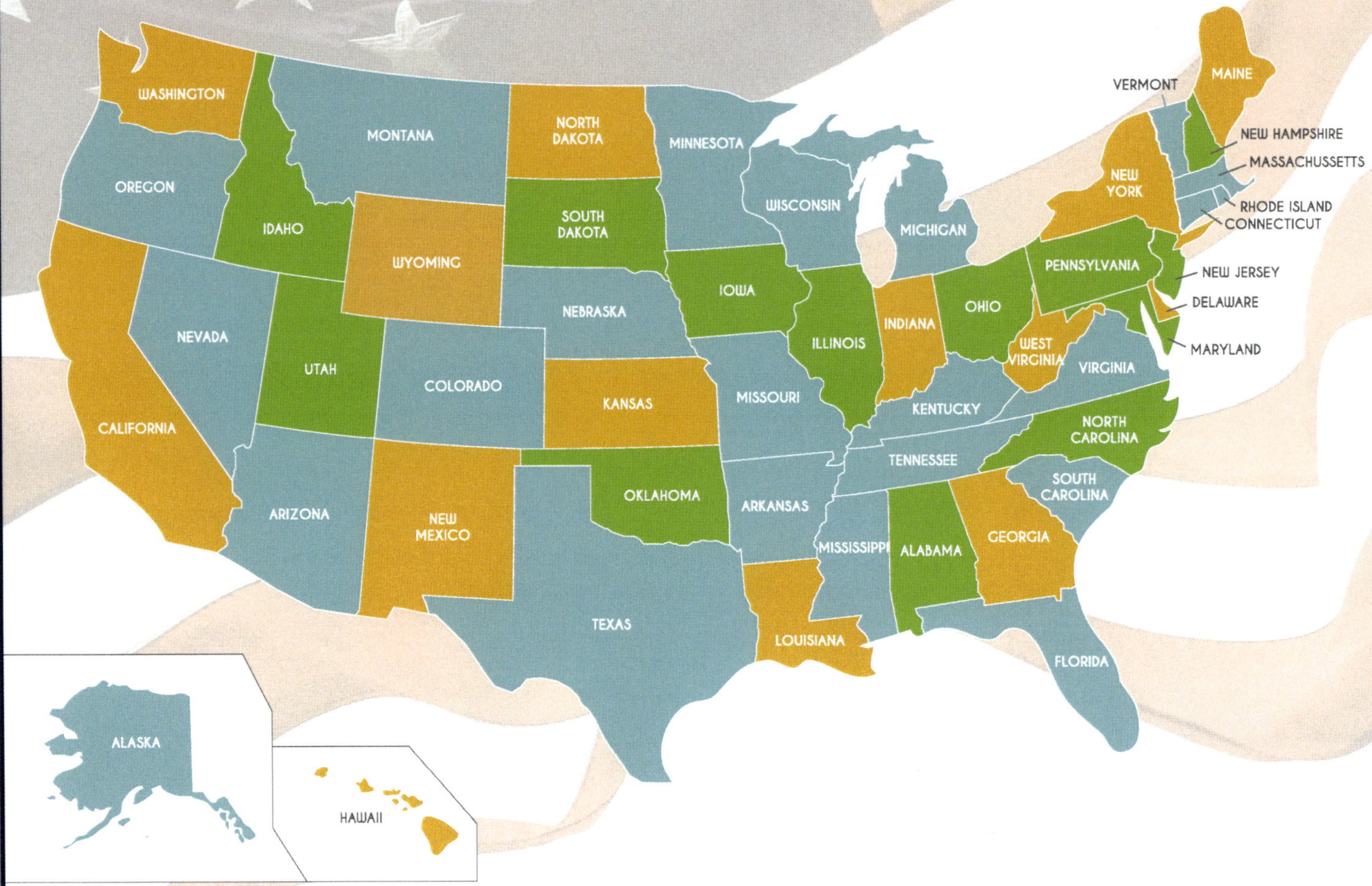

Why Do We Choose Birds as Symbols?

Countries pick birds that represent what matters most to them. When the United States was still just a group of young colonies, leaders chose the Bald Eagle as a powerful symbol of freedom and strength.

As the nation grew and new states joined the union, people wanted to celebrate the amazing birds found in their own regions. That's how the idea of the State Bird was born! Kentucky was the first to make it official in 1926, and other states soon followed.

Each state chose a bird that stood out, whether for its beauty, bravery, usefulness, or popularity with the people. It took many decades for all 50 states to pick their official birds.

Today, the All-Star Bird list is still growing. Some states have added new bird titles like State Raptor, State Duck, and Symbol of Peace or have campaigned to change the ones they picked years ago.

Once a bird gets picked for the All-Star list, it becomes more than just a symbol. The bird gets noticed, talked about, and often protected. People feel proud of "their" bird, learn to spot it, and care more about keeping it safe in the wild.

Let's meet the All-Star Birds that represent us . . .

DID YOU KNOW

Young people—just like you—have helped choose some of these birds! In many states, students voted on which bird should represent them. In some cases, young people even led the charge, campaigning, speaking out, and convincing adults to make their favorite bird an official state symbol.

USA
NATIONAL BIRD

Bald Eagle

Before we explore all the state birds, let's meet the All-Star Bird that represents the whole country: the Bald Eagle!

The Bald Eagle is the USA's official National Bird. It became a national symbol in 1782, just a few years after the country declared independence. That's when it was chosen to appear on the Great Seal of the United States. With its sharp eyes, strong wings, and bold white head, the eagle stood for strength, freedom, and independence, qualities the Founding Fathers hoped would shape the new nation. Today, you can find the Bald Eagle on the President's seal, stamps, military emblems, and even on money.

Bald Eagles live only in North America and are admired across the country. You might see one gliding high in the sky, circling as it searches for fish or soaring over lakes and rivers.

DID YOU KNOW

Even though it's been a symbol for more than 240 years, the Bald Eagle didn't **officially** become the National Bird of the United States until 2024!

LOOK

A giant bird with a white head and tail, chocolate-brown body, and bright yellow beak and feet. Young eagles are all brown and do not get their white head until they're about five years old.

LISTEN

The eagle's call is a series of high-pitched whistles. The sound is much gentler than their fierce look!

LOCATE

They live near lakes, rivers, and coastlines. They mainly eat fish, but also hunt birds, small animals, and will scavenge too.

IOWA
STATE BIRD
NEW JERSEY
STATE BIRD
WASHINGTON
STATE BIRD

American Goldfinch

Now, let's meet the All-Star State Birds, starting from the beginning of the alphabet!

The American Goldfinch is a small, bright yellow bird that adds a splash of sunshine to fields, roadsides, gardens, and backyards. It loves to eat seeds, especially from sunflowers and thistles, and can do some fun tricks to get them—sometimes hanging sideways or even upside down!

Most birds build their nests in spring, but the goldfinch waits until late summer. That's when thistles bloom, and their soft, fluffy seeds make the perfect nest lining.

Three states have selected the American Goldfinch as a State Bird. Iowa was the first, choosing the goldfinch in 1933 because it is common throughout the state and stays all year, even in winter. New Jersey followed in 1935. Washington's path was different, schoolchildren first voted for the Western Meadowlark in 1928, but since that bird was already claimed by many states, later votes favored the goldfinch. In 1951, Washington officially adopted it, making the goldfinch the shared emblem of all three states.

DID YOU KNOW

People from Washington state sometimes call this bird the **Willow Goldfinch** because it's often spotted around rivers and willow trees.

LOOK

Males are bright yellow with black wings and a tiny black cap. Females have softer olive-yellow color with no cap—perfect for blending into the trees.

LISTEN

Their song sounds like a squeaky toy. Listen for "per-chick-o-ree!" or "po-ta-to-chip!" as they fly through the air.

LOCATE

Find them in fields, meadows, and anywhere thistles grow.

CONNECTICUT
STATE BIRD
MICHIGAN
STATE BIRD
WISCONSIN
STATE BIRD

American Robin

With its red-orange belly, cheerful song, and habit of arriving early each spring, the American Robin is one of the most familiar birds in North America. You might see it hopping across lawns, pulling up worms, or perched in a tree eating berries and fruits in fall and winter.

Robins are skilled nest builders. The female makes a sturdy cup-shaped nest from grass, twigs, and mud. Their bright blue eggs are so well-known that the color is called "robin's egg blue." If you want robins to visit your yard, leave a grassy patch for worms, plant berry bushes, or set out a birdbath.

Three states have chosen the American Robin as their State Bird. Michigan was first, in 1931, after a statewide vote crowned it "the best known and best loved of all the birds in the State of Michigan." Connecticut came next in 1943, picking the robin for its abundance in parks, yards, and forests. Wisconsin's story began even earlier. Schoolchildren voted for the robin in 1926–1927, but it didn't become official until 1949.

DID YOU KNOW

Robins use both sight and sound to find worms, watching for movement and listening for sound.

LOOK

A bright reddish-orange chest, gray back, and yellow beak. Males are brighter, while females have softer colors.

LISTEN

It's a sweet, whistling song that sounds like "cheerily, cheer-up, cheerily." Robins are often the very first birds to sing in the morning, even before the sun is up!

LOCATE

Find robins on lawns, gardens, parks, and forest edges, especially after rain.

MARYLAND
STATE
BIRD

Baltimore Oriole

Since 1947, the Baltimore Oriole has been the State Bird of Maryland. Its bold orange-and-black colors match the coat-of-arms of Lord Baltimore's family, who helped start Maryland's first settlements. These birds are special not just for their looks, but also for what they do—they eat lots of bugs! They also weave hanging nests that look like long, dangling socks. To bring them to your yard, try putting out orange slices, grape jelly, or sweet nectar in a special feeder. You can also plant berry bushes or fruit trees to help them feel at home.

LOOK
Males with bright orange bodies, black heads, and white wing bars, and females with yellow-orange undersides and olive backs.

LISTEN
A whistling song that sounds like someone playing a loud flute in the trees.

LOCATE
They live in tall, leafy trees, parks, and quiet neighborhoods, especially during spring and summer.

DID YOU KNOW

The Baltimore Orioles baseball team is named after this colorful bird!

MAINE
STATE
BIRD
MASSACHUSETTS
STATE
BIRD

Black-capped Chickadee

The Black-capped Chickadee is a tiny, lively songbird that stays in New England year-round, even during snowy winters. You can spot it flitting through trees or visiting backyard feeders, often traveling in small flocks. Chickadees are curious and quick, and they take turns keeping watch for danger while the others search for food or explore new branches. A feeder filled with sunflower seeds is a great way to invite chickadees into your yard.

Maine was the first state to choose the Black-capped Chickadee as its State Bird in 1927. This cheerful bird is found all over the state, and its friendly "chick-a-dee-dee-dee" call makes it easy to recognize. Massachusetts followed in 1941, picking the same bird for its own state symbol.

DID YOU KNOW

Chickadees can remember hundreds of hiding spots where they stash food. By fall, their brains **GROW BIGGER** to help them remember all of them!

LOOK

A small puffball with a big round head, black cap and chin, white cheeks, and soft gray wings.

LISTEN

Listen for "chick-a-dee-dee-dee!" The more "dees" at the end, the more excited the bird is.

LOCATE

Look for them in forests, woodsy neighborhoods, parks, and bird feeders—especially ones with sunflower seeds or suet.

LOUISIANA
STATE
BIRD

Brown Pelican

The Brown Pelican is an expert at fishing, diving straight from the sky into the water and scooping up fish with its giant pouch. Large groups of these big, coastal birds live together in places called **rookeries**.

Louisiana has celebrated the pelican for a long time. The bird has appeared on the state flag and seal for over a hundred years. In 1966, it officially became the State Bird, chosen as a sign of coastal pride. At that time, Brown Pelicans were in serious trouble. A harmful pesticide was making their eggs too fragile to survive. Thanks to a major conservation effort, the pelican made an incredible comeback, becoming an even stronger symbol of Louisiana's heritage and resilience.

LOOK

A big bird with a long bill, giant throat pouch, and brownish body. Adults have a creamy yellow head and white neck.

LISTEN

Mostly quiet, but young pelicans squawk loudly when they're hungry!

LOCATE

Check for them in southern coastal beaches and bays, especially around docks and fishing boats.

DID YOU KNOW

One of the most famous Brown Pelican rookeries is in Queen Bess Island in Louisiana's Bartaria Bay.

GEORGIA
STATE
BIRD

Brown Thrasher

The Brown Thrasher is a reddish-brown songbird with a bold personality and an impressive voice. Like its cousin, the Northern Mockingbird, it can sing and mimic the songs of other birds.

On the ground, Brown Thrashers use their long, curved beaks to dig through leaves and soil, tossing debris aside as they hunt for insects, worms, berries, and seeds. They often stay hidden in shrubs or thickets, darting out to grab food before disappearing back into cover.

Georgia first named the Brown Thrasher as its State Bird in 1935, but it didn't become official until 1970.

DID YOU KNOW

Male and female thrashers team up to choose the nest site and build it from twigs, leaves, bark, and grass.

LOOK

A reddish-brown back, streaky white chest, long tail, and bright yellow eyes. Its curved bill is perfect for digging in the dirt.

LISTEN

Listen for musical phrases in pairs. Brown Thrashers often sing a tune and repeat it, mixing bird calls and whistles into one long concert!

LOCATE

Thrashers are found in shrubby yards, forest edges, and gardens, especially leafy places.

ARIZONA
STATE BIRD

Cactus Wren

The Cactus Wren is a desert bird with a BIG personality and an even BIGGER voice! It's the largest wren in North America and is perfectly adapted to life in the Southwest. With bold, spotted feathers and a curious spirit, it's often seen exploring shrubs, poking into crevices, and scolding intruders with its loud "chug-chug-chug" call. This fearless bird builds its nests inside prickly cactus plants, using the sharp spines as protection from predators.

Arizona chose the Cactus Wren as its State Bird in 1931, recognizing it as a tough, resourceful, and lively symbol of the state's rugged desert landscape.

DID YOU KNOW

Cactus Wrens build several extra **"dummy nests"** just to confuse intruders!

LOOK

A brown-and-white speckled bird with a bold white eyebrow, a slightly curved bill, and a long tail often held upright.

LISTEN

Listen for a raspy, chugging "chug-chug-chug" call—some say it sounds like an old sewing machine!

LOCATE

Cactus Wren can be found in dry deserts, especially around cholla and saguaro cacti, prickly bushes, and rocky scrublands.

UTAH
STATE
BIRD

California Gull

More than a coastal scavenger, the California Gull earned its place in Utah's heart thanks to a 1848 event now known as the "Miracle of the Gulls." That year, newly arrived Mormon pioneers watched as hordes of crickets threatened to wipe out their first crops. Just when all seemed lost, thousands of gulls swooped in and devoured the insects, giving the settlers a fighting chance. While the details have become exaggerated over time, the story remains a powerful symbol of resilience, community, and a little divine help for the people of Utah.

In 1955, Utah officially adopted the California Gull as its state bird to honor that story and the role these adaptable birds play in our ecosystem. Whether crunching bugs, fishing along the lake, or even snagging a dropped fry, they prove that not all heroes wear capes—some have feathers!

LOOK

A medium-sized gull with gray wings, a white head and body, yellow legs, and a yellow bill with a small, red and black ring near the tip.

LISTEN

Listen for sharp "yeow" calls and squawky cries, especially near water.

LOCATE

Lakes, rivers, garbage dumps, parking lots, and coastlines—they're super adaptable and love to scavenge.

DID YOU KNOW

The Seagull Monument in Salt Lake City honors the 1848 "Miracle of the Gulls." Built in 1913, it is believed to be the first monument dedicated to birds in the United States.

CALIFORNIA
STATE
BIRD

California Quail

The California Quail is a plump little bird with a fancy head feather called a topknot. Its call sounds like it's saying "Chi-ca-go!" It was chosen as the State Bird of California in 1931 after a big contest where students, schools, and nature clubs all got to vote.

The quail is a gentle, social bird and a familiar sight in both wild spaces and city parks. These birds have adapted to warmer climates by seeking out shade during the hottest parts of the day, foraging in the early morning or evening, or extracting moisture from insects and seeds.

DID YOU KNOW

California Quail can lay huge clutches of up to 16 eggs. Sometimes other females sneak their eggs into the same nest. That sneaky trick is called **"egg-dumping."**

LOOK

A round bird with a curved **"topknot"** feather on its head, scaly belly pattern, and bold black-and-white face (on the male). Females are browner and a bit plainer.

LISTEN

Listen for their distinctive "Chi-ca-go!" call, often repeated in sets of three.

LOCATE

Look for California Quail in scrubby hillsides, grassy fields, backyards, and city parks, often near bushes or under trees.

SOUTH
CAROLINA
STATE
BIRD

Carolina Wren

The Carolina Wren is tiny but mighty! With its warm cinnamon-brown feathers, perky tail, and bold white eyebrow, this little bird has a big, cheerful voice that it uses all year long. You might not spot it right away because it often hides in bushes, trees, and thickets, but once it starts singing, you can't miss it!

Carolina Wrens mostly eat insects and spiders, hopping low to the ground to search under logs and in leaf piles, but will also enjoy seeds and berries in winter. They love cozy, sheltered spaces for nesting and might even set up a home in a hanging flowerpot.

South Carolina chose this bold and friendly bird as its State Bird in 1948, celebrating its year-round presence in the state. In 2000, the state honored the bird by placing it on the South Carolina quarter.

DID YOU KNOW

South Carolina changed its State Bird! The first State Bird of South Carolina was the Northern Mockingbird but was later replaced by the Carolina Wren.

LOOK

A small, reddish-brown bird with a white eyebrow, round body, and a tail often pointed straight up.

LISTEN

Listen for a loud, bubbly song that sounds like "tea-kettle, tea-kettle, tea-kettle!"

LOCATE

Carolina wrens are found anywhere with good hiding spots and insects to eat including wooded backyards, gardens, brush piles, porches, and even mailboxes.

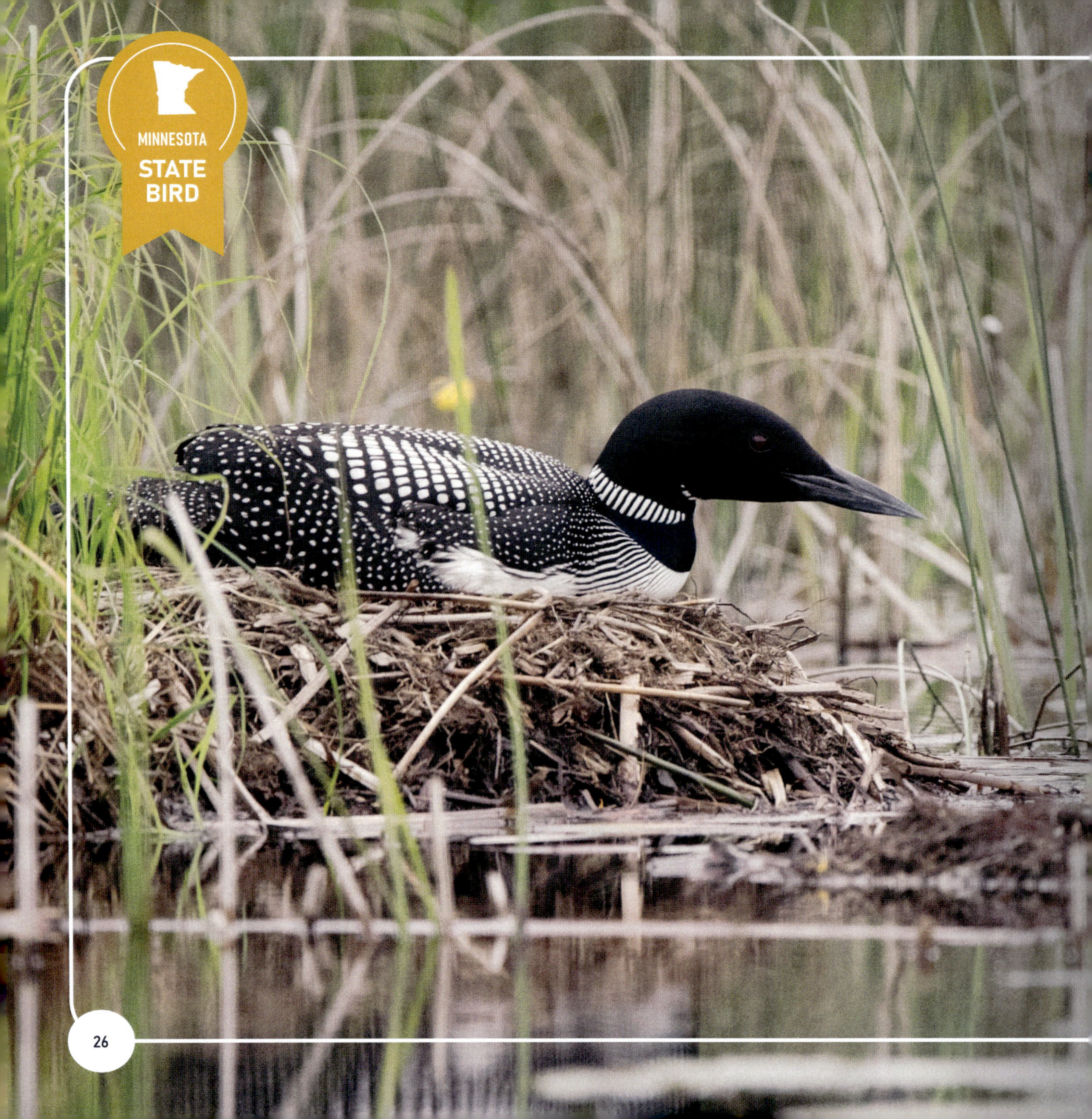
MINNESOTA
STATE
BIRD

Common Loon

The Common Loon is one of Minnesota's most famous lake birds and an amazing swimmer. Its legs are set far back on its body, which makes it clumsy on land but perfect for diving. Loons can slip under the water to chase fish and stay down for up to a full minute. Their spooky, echoing call can carry across an entire lake.

Minnesota made the Common Loon its State Bird in 1961. In 2005, the loon was added to the state quarter. And in 2024, the loon was also added to the state seal, showing just how much this bird means to the people of Minnesota.

DID YOU KNOW

Common Loons can dive up to 200 feet deep and swim underwater like feathered submarines using just their webbed feet!

LOOK

A large black-and-white bird with red eyes, a sharp black bill, and a checkerboard back in summer. In winter, they turn dull gray.

LISTEN

Listen for their haunting yodels and wails—loon calls are some of the most famous sounds in the wild!

LOCATE

Look for loons on northern lakes and large ponds in summer, and along ocean coasts in winter.

DELAWARE
STATE
BIRD

Delaware Blue Hen

The Delaware Blue Hen isn't a wild bird, it's a legendary breed of chicken. During the Revolutionary War, a group of Delaware soldiers led by Captain Jonathan Caldwell brought along fighting chickens said to descend from a famous blue-feathered hen. The soldiers gained a reputation for bravery and toughness in battle, much like their fierce birds. The Blue Hen became a symbol of Delaware pride and was officially named the State Bird in 1939.

LOOK

A chicken with soft blue-gray feathers and a bright red comb.

LISTEN

Like other chickens, Blue Hens cluck, peep, and make a loud "bawk!" when excited.

LOCATE

You might see one on a farm, in a backyard, or at state fairs in Delaware!

DID YOU KNOW

The University of Delaware's mascot, the Fightin' Blue Hen, is named after these legendary birds.

MISSOURI
STATE BIRD
NEW YORK
STATE BIRD

Eastern Bluebird

The Eastern Bluebird is a bright and cheerful bird with a shiny blue back and a rusty orange chest. It loves open fields, orchards, and meadows, where it searches for tasty insects like beetles, caterpillars, and grasshoppers. A bluebird can even spot a bug on the ground from more than 50 feet away!

Missouri was the first state to choose the Eastern Bluebird as its State Bird in 1927. In New York, it took much longer. Back in the 1920s, schoolchildren and bird lovers voted for the bluebird, but lawmakers said no—some even argued that most people in the city only knew pigeons and sparrows! Fans of the bluebird didn't give up, and after more than 40 years, New York finally made it official in 1970.

LOOK

A bright blue back, rusty orange chest, and white belly. Females are softer in color but just as lovely!

LISTEN

Listen for its song that sounds like "cheer, cheerful charmer."

LOCATE

Look for them in nest boxes, open fields, meadows, orchards, and along fences or power lines.

DID YOU KNOW

Eastern Bluebird numbers dropped fast in the 1900s because of habitat loss and invasive birds taking over their nesting spots. But thanks to nest box projects and conservation work, their population has improved.

NEW MEXICO
STATE
BIRD

Greater Roadrunner

The Greater Roadrunner is fast, fierce, and full of desert attitude! It was chosen as the State Bird of New Mexico in 1949 because it's a true desert bird, zipping through the heat on long legs, chasing after bugs, lizards, and small mammals, and chomping on them with a powerful beak. It's a pretty impressive bird in the cuckoo family!

This bird is built for survival in the hot, dry climate of the Southwest. Roadrunners have special nasal glands that help them get rid of excess salt without needing much water. They even reabsorb moisture from their waste, helping them stay hydrated in harsh desert conditions. On chilly mornings, you might catch one sunbathing, lifting its feathers to warm up through a dark patch of skin on its back.

DID YOU KNOW

Roadrunners can run up to 20 miles per hour, fast enough to catch a rattlesnake (and eat it for lunch)!

LOOK

A tall, streaky brown bird with a shaggy crest, a long tail, and legs built for speed. Its head has a patch of bare blue and orange skin behind the eye.

LISTEN

Listen for soft, low "cooing" or a dry clacking sound as it snaps its beak.

LOCATE

Roadrunners are found in deserts, scrublands, and dry open spaces. They are often seen darting across roads or sunning on rocks.

HAWAII
STATE
BIRD

Hawaiian Goose, Nēnē

Say aloha to the Nēnē, Hawaii's very own State Bird! It's the only goose on the All-Star team of state birds and is found nowhere else in the world. With its gentle brown-and-cream feathers and soft "honk," the Nēnē is a rare and special part of Hawaii's natural heritage.

Hawaii made the Nēnē its State Bird in 1957 because it's so unique and closely tied to the islands. Not long ago, it was nearly extinct, with only a few dozen left in the wild. Thanks to years of hard work by scientists and conservation groups, its numbers have slowly grown. Even though the population is still small, wildlife refuges are working hard to protect this amazing bird so future generations can enjoy seeing it and hearing its soft calls across the islands.

LOOK

A gray-brown goose with strong legs, black face, and creamy neck with soft furrows like ripples in sand.

LISTEN

Listen for 'nay-nay'—it's saying its own name!

LOCATE

They can be found on grassy fields, volcanic slopes, and nature reserves on the Big Island, Maui, and Kauai.

DID YOU KNOW

The Nēnē's feet aren't fully webbed like other geese. That makes them better for walking on Hawaii's rocky ground and hardened lava!

VERMONT
STATE
BIRD

Hermit Thrush

The Hermit Thrush is a shy forest songbird with one of the most beautiful voices in North America. Its sweet, flute-like notes echo through Vermont's woodlands in spring and summer, sounding like nature's own lullaby.

Vermont chose the Hermit Thrush as its State Bird in 1941 because of this unforgettable song. You can find it in all 14 counties of the state, usually hopping and scraping through leaf litter or perching low on fallen logs and shrubs. A Hermit Thrush may slowly bob its tail and flick its wings as it forages for food.

DID YOU KNOW

The Hermit Thrush's song has inspired poets and composers. It even shows up in a Walt Whitman poem as a symbol of peace and mystery!

LOOK

A small, brown bird with a reddish tail, spotted chest, and big dark eyes. Often flicks its wings or bobs its tail while perched.

LISTEN

Listen for a soft song that sounds like slow, echoing whistles. It's one of the most loved bird songs in America.

LOCATE

These thrushes are found in cool forests, shady woods, and mountain trails, usually on or near the ground.

Breeding Plumage

Nonbreeding Plumage

Lark Bunting

The Lark Bunting became Colorado's State Bird in 1931 after a teacher named Roy Langdon led the charge, rallying support from students and lawmakers to choose a bird that represented Colorado's prairie, not just its mountains. Schoolkids voted, and the Lark Bunting won the vote! The triumph of the Lark Bunting stands for the wide-open spaces of Colorado and the power of one voice (and one bird) to make a difference.

During breeding season, males sing and dive through the air to impress mates, while females stay camouflaged in the grass. In winter, they form large flocks, sometimes numbering in the thousands during migration and winter.

LOOK

In spring, males are black with white wing patches. In fall and winter, both males and females are streaky brown with pale undersides.

LISTEN

Listen for a rich, warbling song, especially when the male sings in flight above the prairie.

LOCATE

The Lark Bunting can be observed in wide open plains, grasslands, and prairies, usually perched on low shrubs or fences, or hopping through the grass.

DID YOU KNOW

The Lark Bunting is not a lark. It's a type of sparrow that is specialized for life on the prairie.

IDAHO
STATE
BIRD
NEVADA
STATE
BIRD

Mountain Bluebird

The Mountain Bluebird is a bright blue thrush and member of the songbird family. Gentle and graceful, it's often seen in wide-open fields and high country meadows, singing a soft, warbling song as it catches insects in midair. By eating harmful bugs, Mountain Bluebirds help keep nature in balance.

Idaho was the first state to choose the Mountain Bluebird as its State Bird in 1931, after students across the state supported the idea. That same year, Nevada's citizens and schoolchildren also voted for the bird, but it wasn't made official there until 1967—a full 36 years later! Today, this sky-colored bird is a welcome sight in both states' open spaces.

DID YOU KNOW

Female Mountain Bluebirds choose mates who do the best flying tricks in the air.

LOOK

A bright sky-blue bird (males) or a soft gray-blue bird (females) with a round belly and thin beak. Males are especially brilliant in sunlight!

LISTEN

Listen for "chur chur" songs or "few few" calls.

LOCATE

They are often seen perched low, scanning the ground for bugs in open grasslands, high deserts, and mountain meadows in the West.

ILLINOIS STATE BIRD
INDIANA STATE BIRD
KENTUCKY STATE BIRD
N CAROLINA STATE BIRD
OHIO STATE BIRD
VIRGINIA STATE BIRD
WEST VIRGINIA STATE BIRD

Northern Cardinal

The Northern Cardinal is one of the most recognizable birds in America. Its bright red feathers make it stand out in backyards, fields, and forests, especially in winter when everything else looks gray and quiet.

Seven states in the eastern United States have chosen it as their State Bird, since that's where cardinals are common year-round. It all began in 1926 when Kentucky became the first state in the nation to honor the bird as a state symbol. Illinois followed in 1929 after schoolchildren voted for the bird. In 1933, both Indiana and Ohio made it official too. North Carolina joined in 1943, West Virginia in 1949, and Virginia in 1950, making the cardinal the most popular "All-Star" state bird of them all.

Male cardinals are famous for their bright red feathers, while females show off soft tan with touches of red on their wings, tails, and crests. They're loyal mates, caring parents, and happy visitors at backyard feeders, especially if you put out sunflower seeds.

DID YOU KNOW

The male's bright red color comes from what he eats! His feathers get their color from **carotenoids**, natural pigments found in seeds and berries.

LOOK

Males are bright red with a black face and tall crest. Females are warm brown with hints of red and the same spiky crest.

LISTEN

Listen for a sweet whistling song that sounds like "cheer-cheer-cheer" or "birdy-birdy-birdy!"

LOCATE

Cardinals live in forest edges, gardens, and anywhere with trees or thick shrubs.

ALABAMA
STATE
BIRD

Northern Flicker

The Northern Flicker is a type of woodpecker with spotted feathers and a black patch on its chest. There are two kinds of flickers in the United States, the yellow-shafted and the red-shafted. In Alabama, you'll find the yellow-shafted kind. When it flies, you can see flashes of golden yellow!

Because of those golden feathers, people in Alabama started calling the bird the Yellowhammer. Later, during the Civil War, Alabama soldiers wore uniforms with yellow trim. The color reminded people of the Yellowhammer bird, so the soldiers were nicknamed "Yellowhammers." That nickname stuck and eventually Alabama became known as the Yellowhammer State.

In 1927, the Northern Flicker officially became Alabama's state bird. It's the only woodpecker in the country to hold that honor!

DID YOU KNOW

Northern Flickers can eat thousands of ants a day using a tongue that is so long, it wraps around their skull when not in use!

LOOK

A brownish woodpecker with black spots, a black crescent on its chest, and a bright white patch on its rump. Yellow-shafted flickers flash golden wings; red-shafted flickers flash red.

LISTEN

A loud "wick-a-wick-a-wick!" call and hammering on trees—or even on metal signs!

LOCATE

Flickers live in open woods, forest edges, and parks.

ARKANSAS STATE BIRD
FLORIDA STATE BIRD
TENNESSEE STATE BIRD
MISSISSIPPI STATE BIRD
TEXAS STATE BIRD

Northern Mockingbird

The Northern Mockingbird is one of the most popular state birds, claimed by five southern states—Florida, Texas, Arkansas, Tennessee, and Mississippi. This bold, gray-and-white songbird is famous for its amazing voice, copying the songs of other birds and even sounding like squeaky gates or barking dogs. It often sings from the top of a tree or fence, filling the air with music.

Texas was the first state to pick the Northern Mockingbird as its State Bird on January 31, 1927, followed by Florida just a few moths later on April 23, 1927. Arkansas followed in 1929, after the Arkansas Federation of Women's Clubs praised its bug- and weed-eating habits. Tennessee made it official in 1933 after a special vote, and Mississippi joined the states selecting the mockingbird as an All-Star in 1944. Today, this master mimic is a beloved symbol across the South for its lively songs, helpful habits, year-long presence, and fearless personality.

DID YOU KNOW

Some Northern Mockingbirds can mimic up to 200 different sounds!

LOOK
A sleek gray bird with a long tail, white wing patches, and a pale chest. Its tail often flicks as it hops or runs.

LISTEN
Listen for a nonstop mix of copied bird calls, whistles, and chirps. Mockingbirds usually repeat each sound about three times before switching.

LOCATE
They live in yards, parks, and neighborhoods. You may spot them in trees, on fences, or on rooftops where they can sing and spot intruders.

NEW HAMPSHIRE
STATE
BIRD

Purple Finch

The Purple Finch has been the State Bird of New Hampshire since 1957. It's a gentle, friendly bird that brings a splash of beauty to the forests and feeders of the Northeast.

Despite its name, it's not really purple. Males are more raspberry red, while female are more brown and streaky. You'll often see them at feeders in winter, especially if you offer sunflower seeds. They use their strong beaks and tongues to crush seeds and extract the kernel. In winter, they form flocks that can include up to 200 birds, often mixing with other finch species.

LOOK

Males are rosy-red on the head and chest, with brown wings and a slightly chunky beak. Females are brown and white with a bold "eyebrow" and streaky belly.

LISTEN

Listen for a bubbly, warbling song that sounds like a cheerful jumble of notes.

LOCATE

You can find them in woodlands, along forest edges, and at backyard feeders, especially in the northern states.

DID YOU KNOW

New Hampshire's state bird almost wasn't the Purple Finch! In 1957, it beat out the New Hampshire Hen, thanks to strong support and a clever campaign to pass the bill quickly before another state chose it.

RHODE ISLAND
STATE
BIRD

Rhode Island Red

The Rhode Island Red isn't just a chicken, it's a state treasure! It was chosen as the State Bird of Rhode Island in 1954 because it's tough, friendly, and important in farming history. This breed of chicken was developed in Rhode Island and Massachusetts in the 1800s and became famous for being both a great egg layer and a hardy bird that can handle all kinds of weather. Rhode Island Reds lay brown eggs, and they're such good layers that a single hen can produce over 200 eggs a year! It's still a favorite on farms today.

LOOK

A medium-sized chicken with shiny, deep reddish-brown feathers, a bright red comb, and a curious personality.

LISTEN

Listen for soft clucks, peeps, and the classic rooster "cock-a-doodle-doo!"

LOCATE

You can find these birds in barnyards, coops, and even backyard flocks. You might see them scratching in the dirt to find snacks.

DID YOU KNOW

Rhode Island Reds are so important to the state that there's a monument honoring them in Little Compton, Rhode Island, right where the breed began!

SOUTH DAKOTA

STATE BIRD

Ring-necked Pheasant

The Ring-necked Pheasant is not native to the United States, but it is so well known in South Dakota that it was named the State Bird in 1943! These colorful birds were introduced from Asia in the late 1800s and quickly became a favorite across the Great Plains. They are known for their bright feathers, dramatic flight, and tasty game meat. In South Dakota, they are a symbol of outdoor life and wide-open spaces. They're even featured alongside Mount Rushmore on the 2006 South Dakota state quarter!

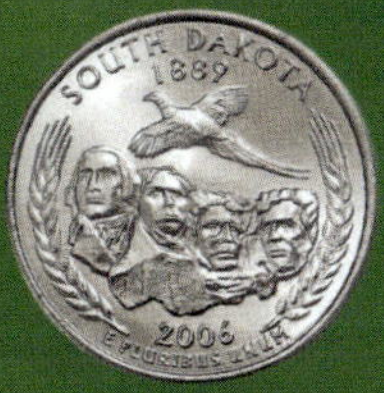

DID YOU KNOW

They can fly 40 miles per hour, but they would rather run than fly when they can!

LOOK

Males are dazzling, with shiny gold and bronze feathers, a white neck ring, a red face, and a long, pointed tail. Females are sandy brown and blend in with grasslands.

LISTEN

Listen for a loud, raspy "koork-kok!" call in spring. When startled, their wings burst into the air with a noisy flap.

LOCATE

They spend most of their time on the ground, feeding on seeds, insects, and grain in open fields, grasslands, and farmlands with tall weeds or crops.

PENNSYLVANIA
STATE
BIRD

Ruffed Grouse

The Ruffed Grouse is a woodland bird that's hard to spot, but once you hear its deep, drumming sound, you'll never forget it. Instead of singing, the male beats his wings rapidly on a log, making a thumping rhythm like a drum solo. Rugged, wild, and perfectly suited to Pennsylvania's forests, this bird has been a state symbol for nearly a century.

The Ruffed Grouse is an All-Star bird in Pennsylvania. Officially, it's the State Game Bird since 1931, but you'll also hear it called the State Bird. As a bird that's hunted seasonally, it represents both the state's wildlife and outdoor traditions.

LOOK

Look for a chicken-sized bird with mottled brown feathers and a short head crest. Males show a black "ruff" of feathers on its neck when displaying.

LISTEN

Listen for the deep "thump-thump-thump" of the male's drumming in spring. It starts slow and speeds up like a lawnmower starting.

LOCATE

Ruffed Grouse live in young, mixed forests, especially where trees are regrowing after cutting or fires.

DID YOU KNOW

Ruffed Grouse are snow specialists! Their feet act like natural snowshoes, helping them walk across snowy ground. In winter, they can bury themselves in the snow to keep warm.

OKLAHOMA
STATE
BIRD

Scissor-tailed Flycatcher

Chosen in 1951, the Scissor-tailed Flycatcher is Oklahoma's State Bird, picked for its beauty, elegance, and love of insects. You can't miss its long, forked tail. It opens and shuts like scissors as the bird twists and dives through the sky catching bugs. Males even put on dramatic "sky dances," soaring up to 100 feet before zigzagging back down to impress females. That amazing tail helps them steer, balance, and show off in midair.

The Scissor-tailed Flycatcher is honored on Oklahoma's 2008 state quarter. On the coin, the bird is flying over a field of native wildflowers representing the state's natural beauty.

DID YOU KNOW

The Scissor-tailed Flycatcher is featured on the crest of FC Tulsa, the state's professional soccer team!

LOOK

A pale gray body with salmon-pink sides and underwings, and an extra-long black-and-white tail that splits like scissors in flight.

LISTEN

Listen for sharp, squeaky "Pik-Pik-Pik" notes and excited chatter, especially during flight.

LOCATE

Look for them in open areas, perched on fences or power lines, feeding on flying insects like grasshoppers, beetles, and wasps.

KANSAS
STATE BIRD
MONTANA
STATE BIRD
NEBRASKA
STATE BIRD
NORTH DAKOTA
STATE BIRD
OREGON
STATE BIRD
WYOMING
STATE BIRD

Western Meadowlark

The Western Meadowlark is the State Bird of six states—Kansas, Montana, Nebraska, North Dakota, Oregon, and Wyoming! With its bright yellow chest, bold black "V," and whistling song, it's a perfect match for the wide-open spaces of the Midwest and West. You can often spot one perched on a fence post in spring, singing loudly to nearby meadowlarks while helping farmers by eating beetles, grasshoppers, and caterpillars.

Schoolchildren helped make the meadowlark a state symbol in Kansas (1937), Montana (1931), and Oregon (1927), where lawmakers followed student votes. Nebraska's 1929 choice came after a campaign by the Nebraska Federation of Women's Clubs and local schools. North Dakota selected it in 1947, later launching the Meadowlark Initiative to protect grasslands for the bird. But Wyoming selected it first, just before Oregon in 1927, after a fourth grader's essay about hearing the bird's song on horseback reached lawmakers and inspired them to act.

DID YOU KNOW

The Western Meadowlark uses strong jaw muscles to pry open soil or bark to access hidden insects, a feeding behavior known as "gaping."

LOOK

A chunky songbird with a bright yellow chest, a black "V" shape on its chest, and brown, streaky wings and back.

LISTEN

Listen for a clear, sweet, whistled song that sounds like a musical greeting from the grasslands.

LOCATE

Western Meadowlarks are found in grasslands, prairies, and open fields.

ALASKA
STATE
BIRD
Winter
Plumage
Summer
Plumage

Willow Ptarmigan

The Willow Ptarmigan (pronounced TAR-mih-gun) is a true Arctic survivor—and it's the State Bird of Alaska! This tough, fluffy bird lives year-round in the wildest, coldest parts of the state. It was chosen in 1955 by schoolkids while Alaska was still a United States territory, and officially adopted as the State Bird when Alaska gained statehood in 1959.

The Willow Ptarmigan is a symbol of toughness and adaptability. A camouflage master, it turns brown in summer and pure white in winter to blend into the tundra. The bird spends more time walking than flying, and even grows feathers on its feet like built-in snow boots to handle the freezing terrain.

LOOK
A plump bird with a small head and feather-covered legs and feet. Males have red eyebrow patches.

LISTEN
Listen for a deep, hoarse clucking or chuckling, often heard before you see them.

LOCATE
Willow Ptarmigans are found in cold climates on the tundra, in willow thickets, and in shrublands.

DID YOU KNOW

In some Arctic cultures, dried ptarmigan poop, known as "urumiit" (pronounced oo-ROO-meet), is considered a delicacy during winter months.

WASHINGTON D.C.
OFFICIAL BIRD

Wood Thrush

Washington, D.C. isn't a state, but it's our Nation's Capital, and it picked an official bird too! In 1967, it chose the Wood Thrush. The Wood Thrush is a bird that doesn't visit backyard birdfeeders. It lives deep in the forest where it's quiet and shady. The bird makes a sweet, musical song that's a favorite among bird lovers. The Wood Thrush is an important part of the forest because it helps control bugs and spreads seeds that grow into new plants and trees.

The Wood Thrush is a powerful traveler. Every year, it flies thousands of miles to spend the winter in Central America, then comes all the way back in the spring.

DID YOU KNOW

The Wood Thrush can sing using both sides of its voice box (**syrinx**) at the same time, as if doing a duet with itself. It's like two birds singing at once!

LOOK

Wood thrushes have a rich reddish-brown color on their head and the back of their neck. Their round belly is white with dark spots, like freckles. They have large eyes that help them see in the shady forests.

LISTEN

It's song typically sounds like "EE-OH-LAY." They often choose a perch with a slope behind them, as if performing on a stage.

LOCATE

Search for the bird in thick leafy forests with moist soil.

GEORGIA
STATE GAME BIRD
MISSOURI
STATE GAME BIRD
TENNESSEE
STATE GAME BIRD
Female
Male

Northern Bobwhite

Now it's time to meet the All-Star Game Birds. These are special birds that are hunted during certain seasons, and play an important role in both wildlife traditions and conservation.

The Northern Bobwhite is a small, round quail with a loud, whistled "bob-WHITE!" call that gives it its name. These birds live in grassy fields, brushy areas, and open woodlands, where they mostly walk or run and only fly when startled. They eat seeds, weeds, and insects. Both the male and female work together to incubate the eggs and care for the chicks.

Georgia named the Northern Bobwhite its State Game Bird in 1970. For many years, the state was known as the "Quail Capital of the World" because wild bobwhites were once common and popular with hunters. Over time, changes to farming and land use caused their numbers to drop, but special conservation efforts are helping them make a comeback in some areas.

Tennessee chose the Northern Bobwhite as its State Game Bird in 1988, celebrating its importance to the state's wildlife and its gentle, familiar call. Missouri followed in 2007, honoring the bobwhite as a symbol of the beauty and wild spirit.

DID YOU KNOW

A group of bobwhites is called a **covey**. They sleep in a circle with their tails in and heads out. They look like a feathery lookout team watching for danger in every direction.

LOOK

A round, chicken-like bird with a brown-and-white face mask, speckled feathers, and short tail. Males have a bold white throat, while females have a buffy tan one.

LISTEN

Listen for its name—"bob-WHITE!" The male calls out "bob-WHITE!" in spring, and the female answers with a sweet four-note whistle.

LOCATE

They live where there are good hiding places such as grassy fields, farms, and woodland edges.

ALABAMA
STATE GAME BIRD
MASSACHUSETTS
STATE GAME BIRD
OKLAHOMA
STATE GAME BIRD
SOUTH CAROLINA
STATE GAME BIRD

Wild Turkey

The Wild Turkey is the largest game bird in North America. These big, bold birds are crafty and hard to catch, making them hunter favorites for centuries. They can run up to 20 miles per hour, fly short distances, and have incredible eyesight. In spring, males fan their tail feathers and "strut" while gobbling loudly to impress mates.

Several states have made the Wild Turkey a symbol. South Carolina was first in 1976, naming it the State Wild Game Bird for its size, beauty, and presence in the forests. Alabama followed in 1980, noting its high turkey population. Oklahoma chose it in 1990, honoring the Rio Grande subspecies that thrives there. Massachusetts added it in 1991 for its cultural and natural history.

Once nearly gone due to overhunting and habitat loss, Wild Turkeys have made a huge comeback. Today, they roam forests, woodlands, and open country as proud state symbols of survival.

DID YOU KNOW

Wild Turkeys are smart. They have great memories, sharp eyesight, and use many sounds to communicate with each other.

LOOK

Turkeys are large brown birds with a fan-shaped tail, bald head, and shiny bronze feathers. Males are larger, puff up, fan their tails, and strut to impress. Turkeys' heads change color with emotion, turning red, white, or blue depending on how excited they are.

LISTEN

Listen for gobbles from the males, plus clucks, purrs, yelps, and other turkey talk!

LOCATE

Look for Wild Turkeys in forests, clearings, and fields.

OREGON
STATE
RAPTOR

Osprey

Now let's meet the All-Star State Raptors. These are powerful birds of prey that rule the skies with sharp talons, keen eyesight, and incredible hunting skills.

The Osprey was chosen as Oregon's State Raptor in 2017 because it's strong, skilled, and perfectly built for life around water. This powerful bird reminds Oregonians of the state's lakes, rivers, streams, and ocean. With its bold markings and graceful flight, the Osprey often makes people stop and stare in wonder.

Ospreys are expert fish hunters, diving feet-first into the water to grab their prey with long, curved talons. They can spot a fish from high above and adjust their dive mid-air to strike with incredible accuracy. Ospreys build large stick nests, often on the tops of trees, poles, or special platforms, and will return to the same nest year after year. Their presence is a sign of healthy waterways, making them a perfect symbol for Oregon's rugged and wild landscapes.

DID YOU KNOW

Ospreys always carry fish head-first through the air. If a fish is caught facing the wrong way, they use their flexible feet to rotate it forward in flight, an incredible skill!

LOOK

A large brown-and-white bird with long wings that bend at the wrist, a dark eye stripe, and a sharp hooked beak.

LISTEN

Listen for a series of high-pitched "kip-kip-kip," kind of like a squeaky toy.

LOCATE

Where there's water, there might be an Osprey! Look for giant nests on top of poles, platforms, or dead trees near lakes, rivers, and coasts.

IDAHO
STATE
RAPTOR

Peregrine Falcon

Peregrines are speedy falcons that mostly hunt other birds. They catch prey in a high-speed dive called a "stoop," grabbing birds right out of the air. A clear extra eyelid works like goggles to protect their eyes while diving. They nest on tall cliffs, city skyscrapers, and even bridges, giving them a bird's-eye view for spotting their next meal.

In 2004, Idaho named the Peregrine Falcon its official State Raptor. Students from two elementary schools helped lead the effort, showing how kids can make a real difference in choosing state symbols.

LOOK

A sleek bird with long, pointed wings, a black head, yellow feet, and a white chest with dark bars.

LISTEN

Listen for sharp, loud "kak-kak-kak!" calls, especially near the nest.

LOCATE

These falcons live on cliffs, coastlines, bridges, and skyscrapers.

DID YOU KNOW

The Peregrine Falcon was featured on Idaho's state quarter in 2007.

OKLAHOMA
STATE RAPTOR

NEW HAMPSHIRE
STATE RAPTOR

Red-tailed Hawk

The Red-tailed Hawk is one of the most common and easy-to-recognize hawks in North America. With its broad wings, reddish-brown tail, and piercing cry, it's a striking sight in the sky. You might see one soaring in wide circles high above or perched on a roadside pole, watching for prey.

Oklahoma made the Red-tailed Hawk its official State Raptor in 2018, and New Hampshire followed in 2019. In both states, schoolchildren helped make it happen. In Oklahoma, the idea started as a 4-H school project. Ten-year-old Ephraim Bowling researched birds of prey, reached out to a state lawmaker, and spoke at the Capitol to explain why the hawk deserved the honor. In New Hampshire, students from Lincoln Akerman School began their campaign in fourth grade, writing letters, speaking at the statehouse, and never giving up until their bill became law.

Red-tailed Hawks soar high above open fields and grasslands, using their sharp eyesight to hunt small animals like mice, rabbits, and snakes. Today, they stand as a symbol of strength, focus, and the natural beauty of the states they represent.

DID YOU KNOW

If you've heard a loud, fierce bird call in a movie, it was probably a Red-tailed Hawk's call!

LOOK

A large hawk with broad wings, a short wide tail (usually red), and a brown body with a pale chest. Young hawks have striped tails instead of red ones.

LISTEN

Listen for a hoarse, screaming "keeeeer!"—it's the classic movie hawk sound!

LOCATE

They live in open fields, deserts, woodlands, and even city parks. You might also see them soaring, looking for mice, squirrels, or rabbits.

MISSISSIPPI
STATE DUCK
SOUTH CAROLINA
STATE DUCK
Male
Female

Wood Duck

Now it's time to meet the State Ducks.

The Wood Duck is one of the most colorful ducks in North America, and it's the official State Duck of both Mississippi and South Carolina. Males are especially striking with shiny green heads, chestnut-colored breasts, and bold white markings. These beautiful birds live in swamps, rivers, and wetlands, where they're often spotted swimming in quiet coves or perching in trees near the water.

Wood Ducks are special because they are one of the few waterfowl that perch and nest in trees. Their sharp nails and webbed feet help them balance on branches and climb into tree cavities or into special nesting boxes placed near ponds and lakes.

Mississippi named the Wood Duck its State Duck in 1974, celebrating its presence across the state's wild, watery places. South Carolina followed in 2009, honoring it as a symbol of the state's rich wetlands and forests.

LOOK

Males are flashy with iridescent green-and-purple head, red eyes, and bold white stripes on its face. Females are soft brown with a white eye ring.

LISTEN

Females give a squeaky "oo-eek!" call. Males give a soft whistle.

LOCATE

These ducks live in swamps, wooded ponds, rivers, and marshes.

DID YOU KNOW

Wood Duck ducklings jump from their nest high in a tree when they are just one day old! They bounce off the ground and run straight for the water, even if the nest is 30 feet up!

WISCONSIN
STATE PEACE SYMBOL

Mourning Dove

The Mourning Dove is a gentle, graceful bird known for its soft cooing sounds. In 1971, Wisconsin named it the State Symbol of Peace, making it the only bird in the United States with that unique title. Despite being a peace symbol, the Mourning Dove is also legally hunted in Wisconsin and several other states (but against the law in other states). The Mourning Dove is the most hunted migratory game bird species in the United States.

The Mourning Dove is a cousin of an extinct bird called the Passenger Pigeon. This bird once filled North American skies in flocks of millions. People believed they were too numerous to ever disappear, but relentless hunting and habitat loss drove the species to extinction by the early 1900s. That loss taught us an important lesson. Even the most common animals can vanish if we're not careful about how we treat them.

DID YOU KNOW

Mourning Dove parents feed their chicks a super-nutritious "crop milk" made in their throats. They are one of the only birds to do this.

LOOK

A slender, grayish-brown bird with a small head, long pointed tail, and black spots on the wings.

LISTEN

Listen for a soft but sad-sounding "coo-OO-oo, oo, oo" that gives the bird its name.

LOCATE

Look for Mourning Doves in open fields, parks, and suburban areas.

The Next All-Star is...YOU!

You have reached the end of the All-Star Bird tour, but the story doesn't stop here!

Now that you've met the birds that soar, sing, strut, and symbolize the USA, you can keep your eyes and ears open for them in the wild.

Maybe you'll spot your own state bird in your backyard, or hear its song while you're at the park. Or maybe you'll remember a fun fact about a bird from another state when you see it on a trip.

These birds aren't just symbols on flags, coins, and license plates. They are living treasures that share our skies, forests, fields, and waters. And every time we protect them and their homes, we are also protecting our own.

So grab your binoculars, keep looking up, and get ready for your own birding adventures.

Because out there, in the air and all around us, America's All-Star Birds are still playing their part in the big, beautiful story of our nation!

Fun Facts about the All-Star Birds

KEY FACTS

The USA has **one** National Bird — the **Bald Eagle**.

Every state has its **own official bird — that's 50 birds!**

8 states picked a **State Game Bird** (hunted seasonally) — Alabama, Georgia, Massachusetts, Missouri, Oklahoma, Pennsylvania, South Carolina, Tennessee.

4 states chose a **State Raptor** (birds of prey) — Idaho, New Hampshire, Oklahoma, and Oregon.

2 states picked **State Ducks** — Mississippi and South Carolina.

Wisconsin is the **only state** with a bird **Symbol of Peace.**

10 states have **two All-Star birds** — Alabama, Georgia, Idaho, Massachusetts, Mississippi, Missouri, New Hampshire, Oregon, Tennessee, Wisconsin.

2 states have **three All-Star birds** — Oklahoma and South Carolina.

BRAG BOARD

1st Place (7 states): Northern Cardinal

2nd Place (6 states): Western Meadowlark

3rd Place (5 states): Northern Mockingbird

4th Place (4 states): Wild Turkey (State Game Bird)

KID POWER!

Young people had a big role in helping pick birds in at least 15 states – Alaska, California, Idaho, Illinois, Kansas, Montana, Nebraska, Nevada, New Hampshire, Oklahoma, Oregon, Washington, West Virginia, Wisconsin, and Wyoming.

The All-Star Bird

Bald Eagle
USA

American Goldfinch
Iowa, New Jersey, Washington

American Robin
Connecticut, Michigan, Wisconsin

Baltimore Oriole
Maryand

Black-capped Chickadee
Maine, Massachusetts

Brown Pelican
Louisiana

Brown Thrasher
Georgia

Cactus Wren
Arizona

California Gull
Utah

California Quail
California

Carolina Wren
South Carolina

Common Loon
Minnesota

Delaware Blue Hen
Delaware

Eastern Bluebird
Missouri, New York

Greater Roadrunner
New Mexico

Hawaiian Goose, Nēnē
Hawaii

Hermit Thrush
Vermont

Lark Bunting
Colorado

Mountain Bluebird
Idaho, Nevada

Northern Cardinal
Illinois, Indiana, Kentucky, North Carolina, Ohio, Virginia, West Virginia

Northern Flicker
Alabama

Hall of Fame

Northern Mockingbird
Arkansas, Florida, Mississippi, Tennessee, Texas

Purple Finch
New Hamshire

Rhode Island Red
Rhode Island

Ring-necked Pheasant
South Dakota

Ruffed Grouse
Pennsylvania

Scissor-tailed Flycatcher
Oklahoma

Western Meadowlark
Kansas, Montana, Nebraska, North Dakota, Oregon, Wyoming

Willow Ptarmigan
Alaska

Wood Thrush
Washington, D.C.

Northern Bobwhite
Georgia, Missouri, Tennessee

Wild Turkey
Alabama, Massachusetts, Oklahoma, South Carolina

Osprey
Oregon

Peregrine Falcon
Idaho

Red-tailed Hawk
Oklahoma, New Hampshire

Wood Duck
Miissippi, South Carolina

Mourning Dove
Wisconsin

All-Star Birds by State

PLACE	SYMBOL	SPECIES	YEAR	PAGE
Alabama	State Bird	Northern Flicker	1927	44-45
	State Game Bird	Wild Turkey	1980	66-67
Alaska	State Bird	Willow Ptarmigan	1955	60-61
Arizona	State Bird	Cactus Wren	1931	18-19
Arkansas	State Bird	Northern Mockingbird	1929	46-47
California	State Bird	California Quail	1931	22-23
Colorado	State Bird	Lark Bunting	1931	38-39
Connecticut	State Bird	American Robin	1943	8-9
Delaware	State Bird	Delaware Blue Hen	1939	28-29
Florida	State Bird	Northern Mockingbird	1927	46-47
Georgia	State Bird	Brown Thrasher	1935	16-17
	State Game Bird	Northern Bobwhite	1970	64-65
Hawaii	State Bird	Hawaiian Goose	1957	34-35
Idaho	State Bird	Mountain Bluebird	1931	40-41
	State Raptor	Peregrine Falcon	2004	70-71
Illinois	State Bird	Northern Cardinal	1929	42-43
Indiana	State Bird	Northern Cardinal	1933	42-43
Iowa	State Bird	American Goldfinch	1933	6-7

PLACE	SYMBOL	SPECIES	YEAR	PAGE
Kansas	State Bird	Western Meadowlark	1937	58-59
Kentucky	State Bird	Northern Cardinal	1926	42-43
Louisiana	State Bird	Brown Pelican	1966	14-15
Maine	State Bird	Black-Capped Chickadee	1927	12-13
Maryland	State Bird	Baltimore Oriole	1947	10-11
Massachusetts	State Bird	Black-Capped Chickadee	1941	12-13
	State Game Bird	Wild Turkey	1991	66-67
Michigan	State Bird	American Robin	1931	8-9
Minnesota	State Bird	Common Loon	1961	26-27
Mississippi	State Bird	Northern Mockingbird	1944	46-47
	State Duck (Waterfowl)	Wood Duck	1974	74-75
Missouri	State Bird	Eastern Bluebird	1927	30-31
	State Game Bird	Northern Bobwhite	2007	64-65
Montana	State Bird	Western Meadowlark	1931	58-59
Nebraska	State Bird	Western Meadowlark	1929	58-59
Nevada	State Bird	Mountain Bluebird	1967	40-41
New Hampshire	State Bird	Purple Finch	1957	48-49
	State Raptor	Red-Tailed Hawk	2019	72-73

All-Star Birds by State

PLACE	SYMBOL	SPECIES	YEAR	PAGE
New Jersey	State Bird	American Goldfinch	1935	6-7
New Mexico	State Bird	Greater Roadrunner	1949	32-33
New York	State Bird	Eastern Bluebird	1970	30-31
North Carolina	State Bird	Northern Cardinal	1943	42-43
North Dakota	State Bird	Western Meadowlark	1947	58-59
Ohio	State Bird	Northern Cardinal	1933	42-43
Oklahoma	State Bird	Scissor-Tailed Flycatcher	1951	56-57
	State Game Bird	Wild Turkey	1990	66-67
	State Raptor	Red-tailed Hawk	2018	72-73
Oregon	State Bird	Western Meadowlark	1927	58-59
	State Raptor	Osprey	2017	68-69
Pennsylvania	State Bird/State Game Bird	Ruffed Grouse	1931	54-55
Rhode Island	State Bird	Rhode Island Red	1954	50-51
South Carolina	State Bird	Carolina Wren	1948	24-25
	State Game Bird	Wild Turkey	1976	66-67
	State Duck	Wood Duck	2009	74-75
South Dakota	State Bird	Ring-Necked Pheasant	1943	52-53

PLACE	SYMBOL	SPECIES	YEAR	PAGE
Tennessee	State Bird	Northern Mockingbird	1933	46-47
	State Game Bird	Northern Bobwhite	1988	64-65
Texas	State Bird	Northern Mockingbird	1927	46-47
United States of America (USA)	National Bird of the USA	Bald Eagle	2024	4-5
Utah	State Bird	California Gull	1955	20-21
Vermont	State Bird	Hermit Thrush	1941	36-37
Virginia	State Bird	Northern Cardinal	1950	42-43
Washington	State Bird	American Goldfinch	1951	6-7
Washington, D.C. (Nation's Capital)	Official Bird	Wood Thrush	1967	62-63
West Virginia	State Bird	Northern Cardinal	1949	42-43
Wisconsin	State Bird	American Robin	1949	8-9
	State Symbol of Peace	Mourning Dove	1971	76-77
Wyoming	State Bird	Meadowlark (Western)	1927	58-59

My All-Star Bird Tracker

America's All-Star birds left their mark on this page—now it's your turn!

Check off the All-Stars you've seen, heard, or learned about!

BIRD NAME	SEEN	HEARD	LEARNED
American Goldfinch	☐	☐	☐
American Robin	☐	☐	☐
Bald Eagle	☐	☐	☐
Baltimore Oriole	☐	☐	☐
Black-capped Chickadee	☐	☐	☐
Brown Pelican	☐	☐	☐
Brown Thrasher	☐	☐	☐
Cactus Wren	☐	☐	☐
California Gull	☐	☐	☐
California Quail	☐	☐	☐
Carolina Wren	☐	☐	☐
Common Loon	☐	☐	☐
Delaware Blue Hen	☐	☐	☐
Eastern Bluebird	☐	☐	☐
Greater Roadrunner	☐	☐	☐
Hawaiian Goose	☐	☐	☐
Hermit Thrush	☐	☐	☐
Lark Bunting	☐	☐	☐
Mountain Bluebird	☐	☐	☐

BIRD NAME	SEEN	HEARD	LEARNED
Mourning Dove	☐	☐	☐
Northern Bobwhite	☐	☐	☐
Northern Cardinal	☐	☐	☐
Northern Flicker	☐	☐	☐
Northern Mockingbird	☐	☐	☐
Osprey	☐	☐	☐
Peregrine Falcon	☐	☐	☐
Purple Finch	☐	☐	☐
Red-tailed Hawk	☐	☐	☐
Rhode Island Red	☐	☐	☐
Ring-necked Pheasant	☐	☐	☐
Ruffed Grouse	☐	☐	☐
Scissor-tailed Flycatcher	☐	☐	☐
Western Meadowlark	☐	☐	☐
Willow Ptarmigan	☐	☐	☐
Wild Turkey	☐	☐	☐
Wood Duck	☐	☐	☐
Wood Thrush	☐	☐	☐

Image Credits

American Goldfinch. Photo by Tina Nord: https://www.pexels.com/photo/yellow-bird-perched-on-yellow-flower-7038102.

American Robin. Photo by Jay Brand. Available at: https://www.pexels.com/photo/american-robin-on-a-lush-green-lawn-31436525.

Bald Eagle. Photo by Lisa Hupp, U.S. Fish and Wildlife Service – Alaska. Public domain. Available at: https://www.flickr.com/photos/usfws_alaska/28149166104.

Baltimore Oriole. Photo by Aaron J Hill. Available at: https://www.pexels.com/photo/close-up-shot-of-a-baltimore-oriole-bird-perched-on-the-branch-12755516.

Black-capped Chickadee. Photo by Aaron J Hill: https://www.pexels.com/photo/close-up-photography-of-a-black-capped-chickadee-on-a-tree-branch-14588461.

Brown Pelican. Photo by Robert So. Available at: https://www.pexels.com/photo/pelican-by-the-sea-19750426.

Brown Thrasher. Photo by A. G. Rosales. Available at: https://www.pexels.com/photo/brown-thrasher-perching-on-birdfeeder-17538196.

Cactus Wren. Photo by Henry. Licensed under CC BY 2.0. Available at: https://www.flickr.com/photos/lakeworth/54509813811.

California Gull, Larus californicus. Photo by Ron Knight. Licensed under CC BY 2.0. Available at: https://www.flickr.com/photos/sussexbirder/20342874102.

California Quail. Photo by Sid Mosdell. Licensed under CC BY 2.0. Available at: https://www.flickr.com/photos/sidm/5268309438.

Carolina Wren. Photo by N. Lewis, Shenandoah National Park, National Park Service. Public domain. Available at: https://www.flickr.com/photos/snpphotos/45449957822.

Common Loon. Photo by John Hanson. Available at: https://www.pexels.com/photo/common-loon-nesting-in-tranquil-wetland-habitat-31347862.

Delaware Blue Hen. Photo by Magali Guimarães: Available at: https://www.pexels.com/photo/close-up-of-a-chicken-26933873.

Eastern Bluebirds. Photo by Kevin Blanzy. Available at: https://www.pexels.com/photo/brown-bird-flying-towards-birdhouse-1156507.

Greater Roadrunner. Photo by Renee Grayson. Licensed under CC BY 2.0. Available at: https://www.flickr.com/photos/132295270@N07/51346520218.

Hawaiian Goose, Nēnē. Photo by Jason Crotty. Licensed under CC BY 2.0. Available at: https://www.flickr.com/photos/46789814@N05/8469357536.

Hermit Thrush. Photo by Christopher Eliot. Licensed under CC BY 2.0. Available at: https://www.flickr.com/photos/eliotc/40901706704.

Lark Bunting, A-9916. Photo by Mark Byzewski. Licensed under CC BY 2.0. Available at: https://www.flickr.com/photos/markbyzewski/53066285545.

Lark Bunting, male, transitional plumage. Photo by Bettina Arrigoni. Licensed under CC BY 2.0. Available at: https://www.flickr.com/photos/barrigoni/25411989353.

Mountain Bluebird. Photo by Tom Koerner/USFWS. Licensed under CC BY 2.0. Available at: https://www.flickr.com/photos/usfwsmtnprairie/47426089782.

Mourning Dove. Photo by Becky Matsubara. Licensed under CC BY 2.0. Available at: https://www.flickr.com/photos/beckymatsubara/41443184325.

Northern Bobwhite. Photo by Steve Maslowski/USFWS. Licensed under CC BY 2.0. Available at: https://www.flickr.com/photos/usfwshq/6862204457.

Northern Cardinal. Photo by Ed McAskill. Available at: https://www.pexels.com/photo/red-cardinal-bird-18814341.

Northern Flicker. Northern Flicker. Photo by A. G. Rosales. Available at https://www.pexels.com/photo/northern-flicker-woodpecker-on-green-grass-32112161/.

Northern Mockingbird. Photo by Andy Morffew. Licensed under CC BY 2.0. Available at: https://www.flickr.com/photos/andymorffew/34831478361.

Osprey. Photo by A. G. Rosales. Available at: https://www.pexels.com/photo/osprey-captures-fish-in-decatur-waters-31634233.

Peregrine Falcon. Photo by Jit Roy. Available at: https://www.pexels.com/photo/peregrine-falcon-in-flight-over-mangalajodi-wetlands-30557613.

Purple Finch. Photo by Skyler Ewing. Available at: https://www.pexels.com/photo/red-fluffy-haemorhous-purpureus-bird-sitting-on-tree-branch-7649227.

Red-tailed Hawk. Photo by Skyler Ewing. Available at: https://www.pexels.com/photo/hawk-flying-in-blue-sky-5748434.

Rhode Island Red Hen. Photo by Garrett Heath. Licensed under CC BY 2.0. Available at: https://www.flickr.com/photos/garrettheath/9340819782.

Ring-necked Pheasant. Photo by Graeme Travers. Available at: https://www.pexels.com/photo/pheasant-on-grass-18866471.

Ruffed Grouse. Photo by Aaron J Hill. Available at: https://www.pexels.com/photo/brown-bird-on-tree-branch-5977729.

Scissor-tailed Flycatcher. Photo by Larry Smith. Licensed under CC BY 2.0. Available at: https://www.flickr.com/photos/cachesmith/53728152874.

Western Meadowlark. Photo by Elijah Pilchard. Available at: https://www.pexels.com/photo/western-meadowlark-perched-on-a-wooden-log-12792201.

Willow Ptarmigan (Summer Plumage). Photo by Þorsteinn Friðriksson. Available at: https://www.pexels.com/photo/a-bird-standing-on-a-wood-8613206.

Willow Ptarmigan (Winter Plumage). Photo by Denitsa Kireva. Available at: https://www.pexels.com/photo/close-up-of-white-bird-16752483.

Wild Turkey. Photo by Mohan Nannapaneni. Available at: https://www.pexels.com/photo/wild-turkey-in-nature-16956763.

Wood Ducks. Photo by Mike's Birds. Licensed under CC BY-SA 2.0. Available at: https://www.flickr.com/photos/pazzani/32824040474.

Wood Thrush. Photo by Nate Rathbun/USFWS. Public domain. Available at: https://www.flickr.com/photos/usfwsmidwest/34499552330.

For Parents and Teachers

The facts for this book are drawn from trusted scientists, historians, bird experts, and nature organizations. Here are some of them:

- American Bird Conservancy
- Field guides and nature books
- State government websites
- State symbols USA
- The Audubon Society
- The Cornell Lab of Ornithology
- U.S. Fish and Wildlife Service

If you want to explore more about birds, check out:

- The Young Birder's Guide to Birds of North America
- National Geographic Kids Bird Guide of North America

Acknowledgements

I am deeply grateful to the many people whose support made this book possible. My heartfelt thanks go to DeWitt Allen, Doe Boyle, Gilles Carter, Ricci Cummings, Kimberly Behre Kenna, Mark Seth Lender, Patrick J. Lynch, Florence McBride, Tom Stephenson, Jana Rade, Susan Wade, and David Waters for their generosity, encouragement, and belief in this project.

About the Author

Gail Martino is a writer, corporate consultant, former R&D scientist, and birder based in New Haven, Connecticut. She serves on the boards of Connecticut Audubon and the New Haven Bird Club. With the New Haven Bird Club, she has led bird walks, organized speaker programs, and co-edited Chickadee Tales: A New Haven Bird Club Anthology. Gail has also served as a citizen scientist for multiple bird conservation projects across Connecticut. A former educator, she taught for the Johns Hopkins Center for Talented Youth and later served as an assistant professor at Colgate University. She loves inspiring nature lovers of all ages to discover the magic of birds.

Made in the USA
Coppell, TX
05 November 2026

11315629R00059